SLEEP

noun
The natural state of rest during which your eyes are closed and you become unconscious.

Barbara Herkert

illustrated by
Daniel Long

Albert Whitman & Company
Chicago, Illinois

Sleep.

Our bodies slow down.

Our muscles relax.

Our senses—touch and sound and smell and taste—relax too.

Our eyes close.

All day long, our bodies and brains work hard. During sleep, our bodies restore themselves. Our brains need less energy when we sleep deeply. We go to a subconscious place to focus on feelings, thoughts, and memories outside of our awareness.

Animals need sleep or periods of rest too. How long they sleep, how deeply they sleep, what happens in their brains—is as varied as they are. Nature adapts sleep to different habitats and lifestyles.

Each night people have two types of sleep: REM, or sleep with rapid eye movement, and non-REM, sleep without rapid eye movement.

REM sleep is usually when we dream, but people can also dream in non-REM sleep. Mammals and birds show signs of REM, which means they probably dream too. Reptiles and other cold-blooded animals don't appear to dream.

During REM sleep, the brain appears to be awake, active, and filled with images. During non-REM sleep, the brain appears to be calm. Scientists know this from studying human brain waves with a test called an EEG.

There are many ideas about why and how we dream. To find answers, scientists study different parts of the brain that are active during dreaming.

Although it's not clear why we dream, researchers know we need to dream to stay healthy. When we don't get REM sleep, we can get diseases and memory problems more easily.

We don't know why or what animals dream.

When it gets dark, our brains tell the retinas in our eyes to stop sending signals. Our brains then tell our bodies to produce melatonin, a hormone that makes us drowsy, and they also tell our bodies that it is time to tune out the external world. Our brains go into a subconscious state.

There are five stages of sleep. The first, non-REM sleep when we begin to fall asleep, lasts a few minutes. The second is a non-REM light sleep that lasts about twenty-five minutes. The third and fourth stages of non-REM sleep are the deepest stages of sleep. The fifth stage is REM sleep, about ninety minutes after we fall asleep.

These stages repeat in a cycle all night. The REM stage lasts about ten minutes in the first cycle and can increase to about sixty minutes by the last cycle of a full night's sleep.

The ways humans and most animals sleep are very different from each other. Humans, as well as apes and chimpanzees, mostly sleep in one long stretch.

Elephants and giraffes sleep in short bursts, with several cycles of sleep throughout the day.

Frigate birds and dolphins rest with one half of their brain while the other half stays aware of danger.

Bears and bats hibernate during cold winter months to survive when food is scarce. Their body temperatures drop. Their breathing, heart rates, and body functions slow down to save energy.

Sharks have active and restful periods throughout the day. Some scientists believe that sharks don't sleep, but only rest, because many sharks move continuously. Scientists aren't certain how long or when sharks rest or sleep.

Two dozen species of sharks, including great white sharks and hammerheads, need to swim while they rest. Swimming keeps water flowing through their gills, allowing them to breathe the oxygen in the water.

Some bottom-dwelling sharks have small openings called spiracles behind each eye. Spiracles force water across these sharks' gills while they rest at the bottom of the sea. Other bottom-dwelling sharks draw water into their mouths and over their gills.

Sharks that swim while they rest include great white sharks, hammerheads, megamouths, makos, whale sharks, and thresher sharks. Many live in the open ocean in temperate climates.

Half a shark's brain shows sleep brain waves while the other half stays active. The sides alternate rest and activity so part of a shark's brain is always rested. Scientists aren't certain how often the two sides alternate. Many experts believe the pattern changes throughout the night.

Nurse sharks, whitetip reef sharks, Caribbean reef sharks, wobbegongs, and lemon sharks have spiracles and rest while lying on the ocean floor. This way of resting suits their bottom-dwelling lifestyles. They often live in tropical coral reefs.

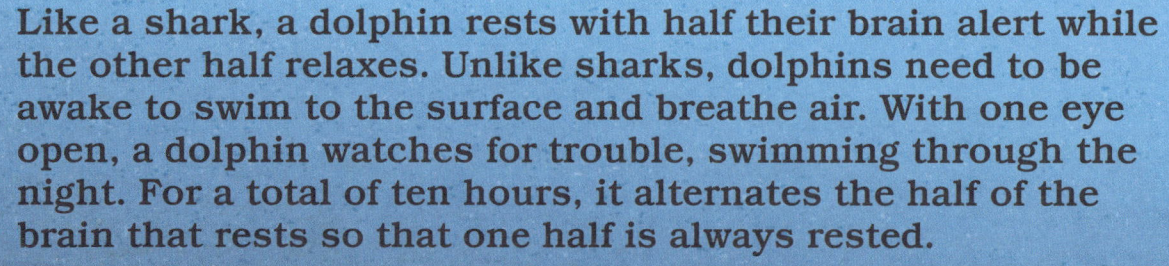

Like a shark, a dolphin rests with half their brain alert while the other half relaxes. Unlike sharks, dolphins need to be awake to swim to the surface and breathe air. With one eye open, a dolphin watches for trouble, swimming through the night. For a total of ten hours, it alternates the half of the brain that rests so that one half is always rested.

Unlike other mammals, dolphins, whales, and porpoises don't show signs of REM sleep. During REM sleep, muscles are temporarily paralyzed, so dolphins and other cetaceans would sink to the bottom of the ocean and drown.

In its first month, a baby dolphin rests against its mother, who moves constantly to reduce danger from predators and also to help regulate her baby's body termperature.

Gray whales and humpbacks sleep lying near the surface. This is called logging because they look like floating logs.

Sperm whales look like standing giants when they sleep, floating tail-down in groups called drift dives. Scientists do not know why they sleep this way.

Whales sleep in short bursts of ten to fifteen minutes at a time because they need to be awake to breathe. They also need to move to stay warm.

Sperm whales sleep for short sessions totaling about one and a half hours a day. Gray whales sleep up to 10 hours a day.

Wrapped snuggly in kelp beds on the surface of the water, sea otters reach for each other's paws before they sleep.

Baby otters sleep on their mothers' bellies. With kelp to anchor them in place in the rolling sea, sea otters sleep soundly in groups called rafts, without fear of drifting. Sea otter families stay together in this way, floating serenely on their backs.

Frigate birds sleep while flying. Sometimes they shut down half their brains while the other half stays mindful, with one eye open to avoid midair collisions. Sometimes they sleep with both halves of their brains. Scientists have discovered that for a few seconds each night, the birds experience REM dream sleep.

Frigate birds sleep for a total of 42 minutes over 24 hours.

While they are over water, frigate birds sleep in the early evening, shortly after sunset.

When frigate birds experience dream sleep for a few seconds while in flight, they fall in altitude as their muscles rest.

After flying and feeding on squid and flying fish for up to two months at a time, frigate birds nest on land in mangrove trees and scrubby bushes, where they sleep in one-minute periods for a total of twelve hours a day.

Just two hours is all the sleep an elephant requires during one day, snatched in short spurts through the night.

Elephants spend most of the day eating—six hundred pounds of grass, roots, and bark. When they are full, their brains send signals to rest.

Elephants usually sleep standing up. It takes a lot of time and energy to get up from the ground, and they need to respond quickly to predators such as lions and hyenas, or even crocodiles. Baby elephants lie down under their mothers' watchful eyes.

Elephants can chew their food while sleeping. They also snore and moan while asleep. It's not certain if this means they are dreaming.

Elephants experience REM dream sleep every few days for several seconds. They have to lie down for REM sleep because their muscles become temporarily paralyzed, making them unable to remain upright.

Elephants often take turns sleeping to keep each other safe. They often sleep leaning against trees.

Giraffes sleep for only thirty minutes a day, sometimes lying down, sometimes standing. With their long legs, it's slow and awkward for them to get up from the ground. They need to be ready to respond to danger—lions, leopards, and hyenas.

A short sleep cycle leaves plenty of awake time for foraging and chewing cud.

Baby giraffes lie down with their legs tucked under their bodies and their heads resting on rump pillows.

Adult giraffes rarely sleep for more than 5 minutes at a time, sometimes keeping one eye open to watch for danger.

Giraffes and other herbivores have joints in their legs that "lock" so they can sleep standing up without falling over.

Koalas are awake at night and asleep during the day. It takes a lot of sleep to digest a koala's leafy diet—up to twenty hours of sleep throughout the day. Koalas hunker down in eucalyptus trees to sleep, hanging on firmly with clawed fingers and toes.

Baby koalas, called joeys, sleep in their mothers' pouches until they are 6 or 7 months old, when they sleep on their mothers' backs or bellies.

Koalas need more sleep than most animals because the eucalyptus leaves they eat contain toxins that take a large amount of energy to digest. Sleeping helps conserve their energy.

Bats hang upside down to sleep in crevices, caves, trees, and buildings, gripping rock ceilings or branches with their claws. This upside-down sleeping puts them in a good position for a quick flight takeoff in case of predators, such as large owls. Bats sleep in colonies of one hundred bats or more, sometimes waking for a few minutes to groom themselves and each other or to socialize.

When it starts to get dark, bats drop and flap away from roosts—the places where they gather to sleep—swooping and gobbling nighttime insects such as mosquitoes and moths. Bats rest high in trees or buildings for a moment or two, then flap off to hunt again. When nighttime bugs disappear at daybreak, bats return to their roosts.

Special tendons in a bat's toes and claws "lock" in place while the bat hangs upside down.

Bats hibernate during winter in cold climates when food is more difficult to find. Their body temperatures drop, and their body functions slow down.

Cicadas that are on a 17-year cycle wait for the perfect soil temperature—64°F—before journeying to the surface. Then, all the cicadas in the area use a newly made exit hole from the earth to molt and mate.

Torpor is a type of rest in which the cicada's body remains motionless and stops functioning. This is different from hibernation, when the body still functions but breathing and heart rates slow and body temperature drops.

Cicadas rest underground when they are not busy sipping root sap or tunneling through the earth. They go into torpor when the temperature drops to between 32 degrees Fahrenheit and 50 degrees Fahrenheit. Their bodies stop functioning until the weather warms again.

Out of thousands of different cicada species in the world, only three species emerge from the ground every seventeen years. They don't spend those years hibernating underground. They are active, drinking sap, making tunnels, and growing in their nymph forms.

Other species of cicadas emerge every year, or every two to five years, or every thirteen years. Scientists aren't certain why there are different time periods.

Because orb-weaving spiders are on a 17-hour internal cycle, they become inactive at a different time each day, and they wake to get to work at a different time each night.

Scientists are not sure why orb-weaving spiders have this unusual internal clock. No other animal shows this offbeat timing.

The internal clocks of orb-weaving spiders alert them to wake and repair their webs by night and rest during the day. These spiders become active around dusk, working through the night on their webs until a few hours before dawn.

During the day, orb-weaving spiders sit motionless in their webs to rest, moving only when prey, such as flies and gnats, get stuck in their webs.

A bullfrog shuts its eyes to rest by closing one transparent eyelid and one semi-transparent eyelid, while a third transparent eyelid stays open. Its eyes sink down into its mouth for protection. Bullfrogs always stay alert and ready to react to predators such as herons, water snakes, and raccoons.

In the winter bullfrogs hibernate underwater, at the bottom of a lake or pond. Mucus glands in their skin absorb oxygen from the water, allowing them to breathe. They use their bodies' energy stores from food eaten in warmer weather to survive the cold.

Bullfrogs are active both day and night. Some scientists think these frogs might have a way of resting while awake.

In the spring, a bullfrog's body temperature rises. The bullfrog darts to the water's surface, ready to snag a meal of worms, insects, fish, snakes, or turtles.

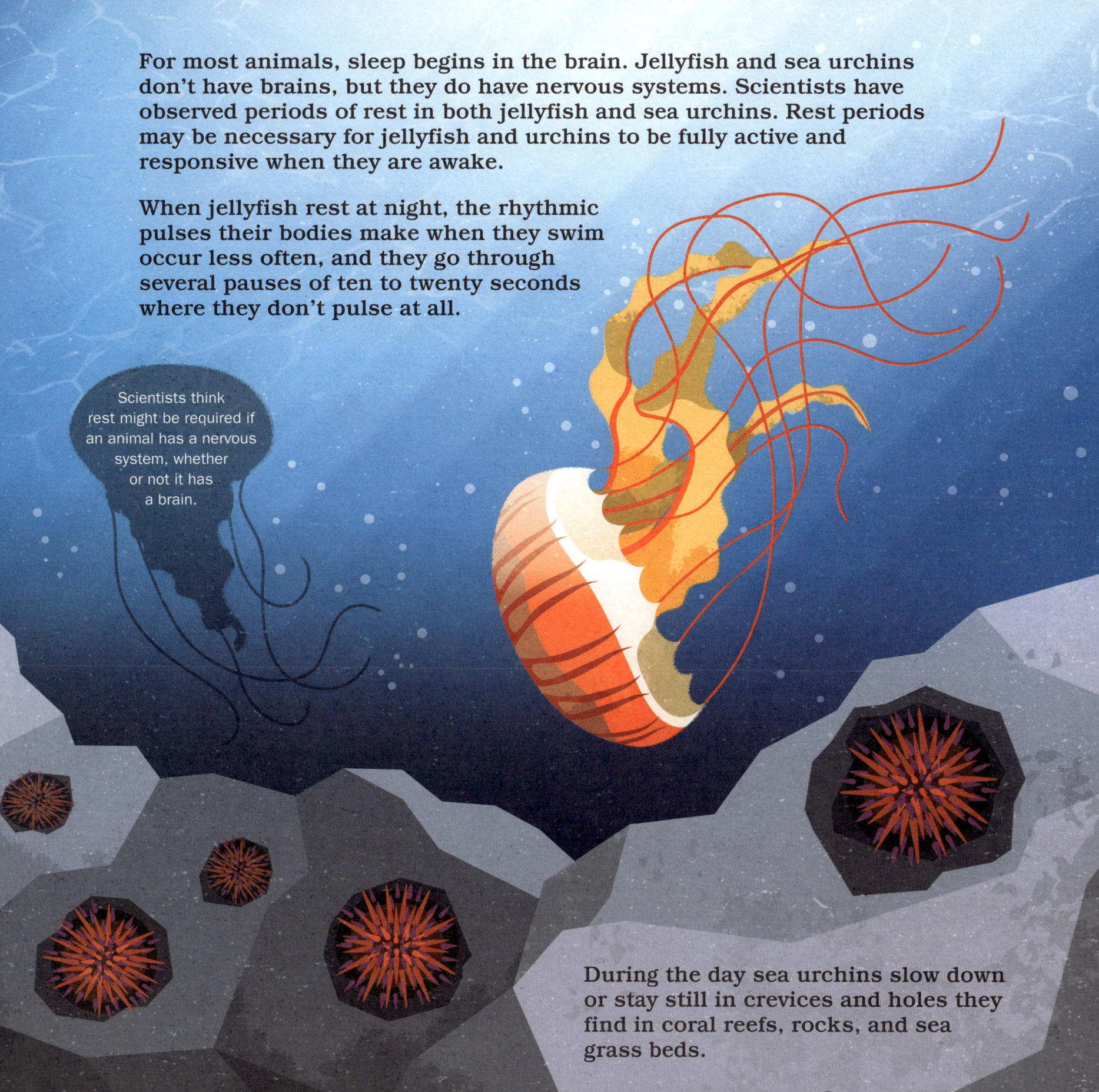

For most animals, sleep begins in the brain. Jellyfish and sea urchins don't have brains, but they do have nervous systems. Scientists have observed periods of rest in both jellyfish and sea urchins. Rest periods may be necessary for jellyfish and urchins to be fully active and responsive when they are awake.

When jellyfish rest at night, the rhythmic pulses their bodies make when they swim occur less often, and they go through several pauses of ten to twenty seconds where they don't pulse at all.

Scientists think rest might be required if an animal has a nervous system, whether or not it has a brain.

During the day sea urchins slow down or stay still in crevices and holes they find in coral reefs, rocks, and sea grass beds.

Babies need sixteen hours of sleep a day to develop and grow.
Children require nine and a half hours.
Adults should sleep seven hours.

When babies and children sleep, growth hormone is released.
Proteins that our bodies rely on to fight infections are produced.
Our brains develop.

As we grow, we require less sleep hours for growing,
but we still need sleep to clear our brains.
When we get enough sleep, we awake with new ideas.

Many scientists think that every animal needs some kind
of rest or sleep to restore its body and function at its best.

Author's Note

In writing this book about sleep, I started looking at the animals and insects around me with renewed fascination. It seems that every creature needs some type of rest or sleep in order to thrive. Each has adapted to a type of rest or sleep that best suits where it lives, and its lifestyle. Nature assures that whether you sleep in water, on land, or in the air, there is a way for you to get the sleep or rest your body needs.

Resources

books for children

Na, Il Sung. *A Book of Sleep*. New York: Alfred A. Knopf, 2009.

Peyrols, Sylvaine. *How Animals Sleep*. Abingdon, UK: Moonlight Publishing, 2015.

Prendergast, Kate. *Sleep: How Nature Gets Its Rest*. Somerville, MA: Candlewick Press, 2019.

Glossary

cud: partly digested food.

EEG: a medical test that detects electrical activity in the brain.

energy stores: energy from food that is stored throughout the body for later use.

gravity: an invisible force that pulls everything toward Earth.

growth hormone: a chemical our bodies produce that spurs growth in children and adolescents.

herbivores: animals that mainly eat plants.

hibernate: to enter a state of inactivity, low body temperature, slow breathing, and slow heart rate when there is a long period of cold.

hormone: a chemical messenger sent by the brain to travel through the blood and help regulate bodily functions.

kelp: a type of large, brown seaweed that grows in shallow ocean water near coasts.

lens: transparent, flexible tissue in the eye that helps focus light and images on the retina.

Glossary, continued

mammals: warm-blooded animals that produce milk for their young, are typically born live, and that have hair or fur and backbones.

melatonin: a hormone produced in the brain that helps the human body know when it's time to sleep and when it's time to wake up.

molt: to shed old feathers, hair, skin, or an old shell to make way for new growth. Cicadas molt by shedding their shells and unfurling their wings.

mucus glands: organs in the body that help maintain the correct balance of water. They produce mucus that gives protection from bacteria by killing and disabling germs.

nervous system: a network of neurons and cells that carry messages to and from the brain, spinal cord, and various parts of the body.

non-REM (non-rapid eye movement) sleep: when breathing and heart rate are slow and regular, blood pressure is low, and the sleeper is mostly still.

nymph: an immature form of an insect that is usually wingless.

predator: an animal that hunts and eats other animals.

proteins: substances in the body that are essential to all living organisms.

REM (rapid eye movement) sleep: the stage of sleep when dreams happen, with irregular breathing, irregular heart rate, and involuntary muscle jerks.

reptiles: cold-blooded animals that lay eggs on land and have backbones or spinal columns, and dry skin covered with scales or horny plates.

retina: a layer of tissue at the back of the eye that receives light from the lens, converts it to signals, and transmits them to the brain.

semi-transparent: partially see-through.

subconscious: a part of the mind that is not fully aware of our actions and feelings but influences them.

temperate climates: patterns of weather that have wider temperature ranges throughout the year and more distinct seasonal changes compared to tropical climates.

tendons: pieces of connective tissue that attach muscle to bone.

torpor: a state of stillness and low body temperature during periods when body temperature drops to between 32°F and 50°F, allowing an animal or insect to conserve energy.

toxins: poisons produced by plants or animals.

transparent: completely see-through.

unconscious: a part of the mind containing feelings, thoughts, urges, and memories that we are not aware of.

Many thanks to biologist and author
Melissa Denny for sharing her knowledge
and many fascinating conversations.
And to little sleepers everywhere.—BH

For Hannah, whom I would
be lost without—DL

Library of Congress Cataloging-in-Publication data is on file with the publisher.
Text copyright © 2022 by Barbara Herkert
Illustrations copyright © 2022 by Albert Whitman & Company
Illustrations by Daniel Long
First published in the United States of America in 2022 by Albert Whitman & Company

ISBN 978-0-8075-7435-5 (hardcover)
ISBN 978-0-8075-7436-2 (ebook)

All rights reserved. No part of this book may be reproduced or transmitted in any form
or by any means, electronic or mechanical, including photocopying, recording, or by any
information storage and retrieval system, without permission in writing from the publisher.

Printed in China
10 9 8 7 6 5 4 3 2 1 WKT 26 25 24 23 22

Design by Aphelandra

For more information about Albert Whitman & Company,
visit our website at www.albertwhitman.com.